ACTION MATH

PATTERNS

Ivan Bulloch

Consultants
Wendy and David Clemson

WORLD BOOK

in association with

TWO-CAN

2 Looking at Patterns

There are patterns all around us. You can find them in sidewalks, in the stitches of your sweater, and on a butterfly's wings. A pattern is made when shapes or numbers are put in a sequence and repeated. Look around you. How many different patterns can you see?

We use patterns to help us make sense of the world. Math is all about patterns. The activities in this book will help you:
- sort things into groups.
- match similar things.
- find out how things fit together.
- create patterns, shapes, and designs.

Make a patterned snake to hang from your ceiling.

Painting a Spiral

● Draw a spiral on a piece of poster board. Start from the edge of the board and gradually spiral in toward the center. You may need to draw a few spirals for practice first.

● Cut out the snake starting from the end of the spiral on the edge of the poster board.

● Paint a snake pattern on your spiral, or decorate it with colored paper.

● Ask an adult to string a piece of thread through the middle of the spiral. Now hang up your snake.

Here's what you learn:
● how to create repeating patterns.
● how to change a flat shape into a three-dimensional one.

6 Beads

Threading beads is a good way to make a pattern. Look around your house or school for things to use as beads. Here are some ideas for making your own beads.

Paper Beads

● Glue together two pieces of colored paper. Tear out a triangle shape. Roll the shape around a pencil and tape down the narrow end.

● Make a simple paper bead with a long strip of colored paper. Roll it around a pencil, then tape down the end. You could decorate the paper before making your beads.

Pasta Beads

● Pick pasta shapes with holes through the middle.
● Paint the shapes and let them dry.

Clay Beads

● Use the type of modeling clay that dries overnight to make these beads.
● Make small balls of clay and ask an adult to make a hole in each with a knitting needle or toothpick. Let the beads harden.

Sort Them Out

How many different types of beads have you made or collected?
● Sort the beads into different colors and shapes.

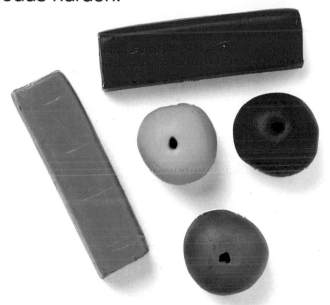

Here's what you learn:
● how to sort things into different groups or categories.
● how to match similar things.

Beads to Find

If you look, you should be able to find lots of things to use as beads. We used straws, plastic beads, and even peanuts. Can you think of anything else you could use?

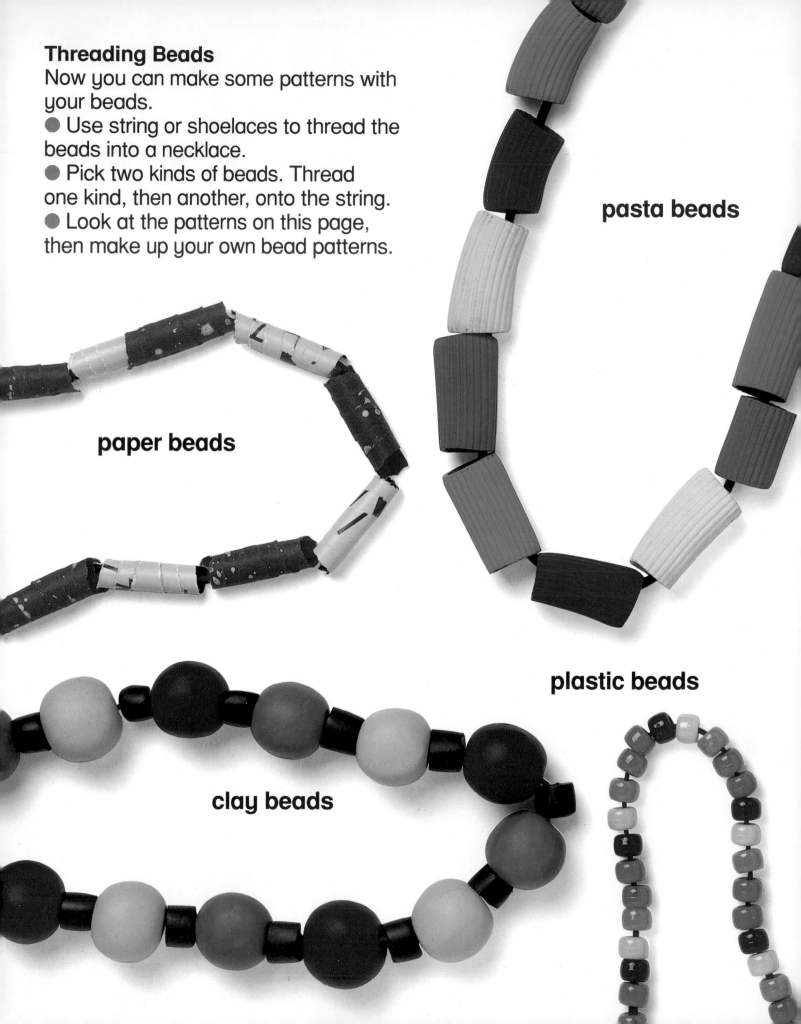

Threading Beads

Now you can make some patterns with your beads.

● Use string or shoelaces to thread the beads into a necklace.

● Pick two kinds of beads. Thread one kind, then another, onto the string.

● Look at the patterns on this page, then make up your own bead patterns.

pasta beads

paper beads

plastic beads

clay beads

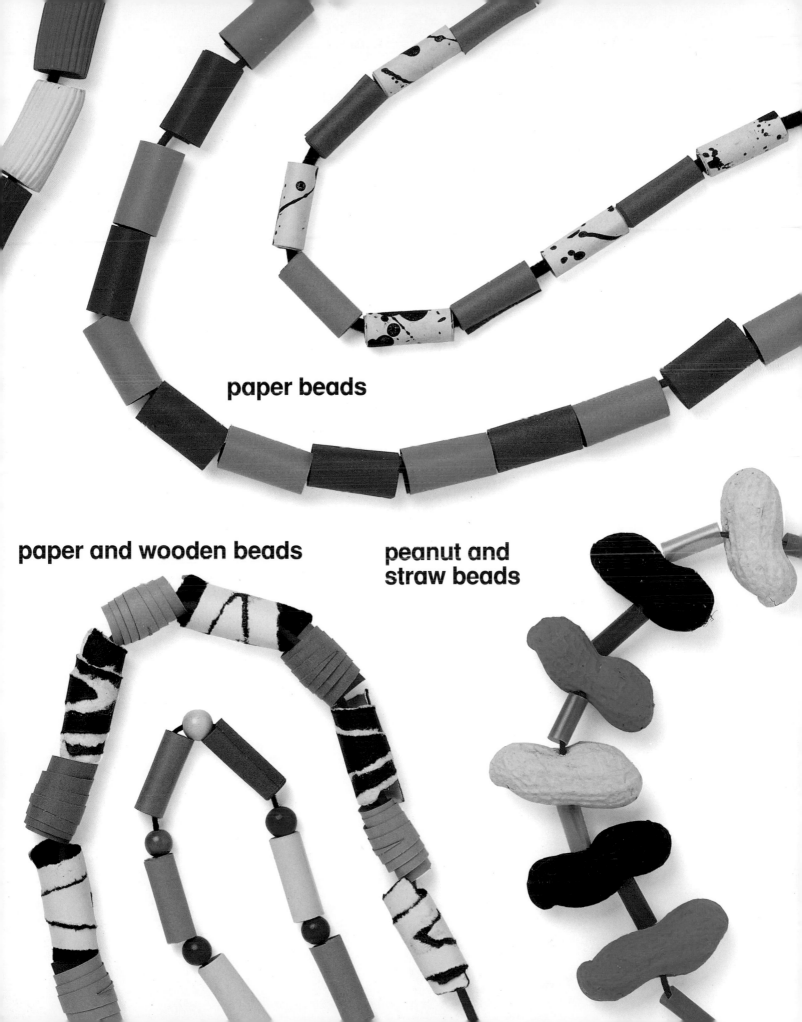

paper beads

paper and wooden beads

peanut and
straw beads

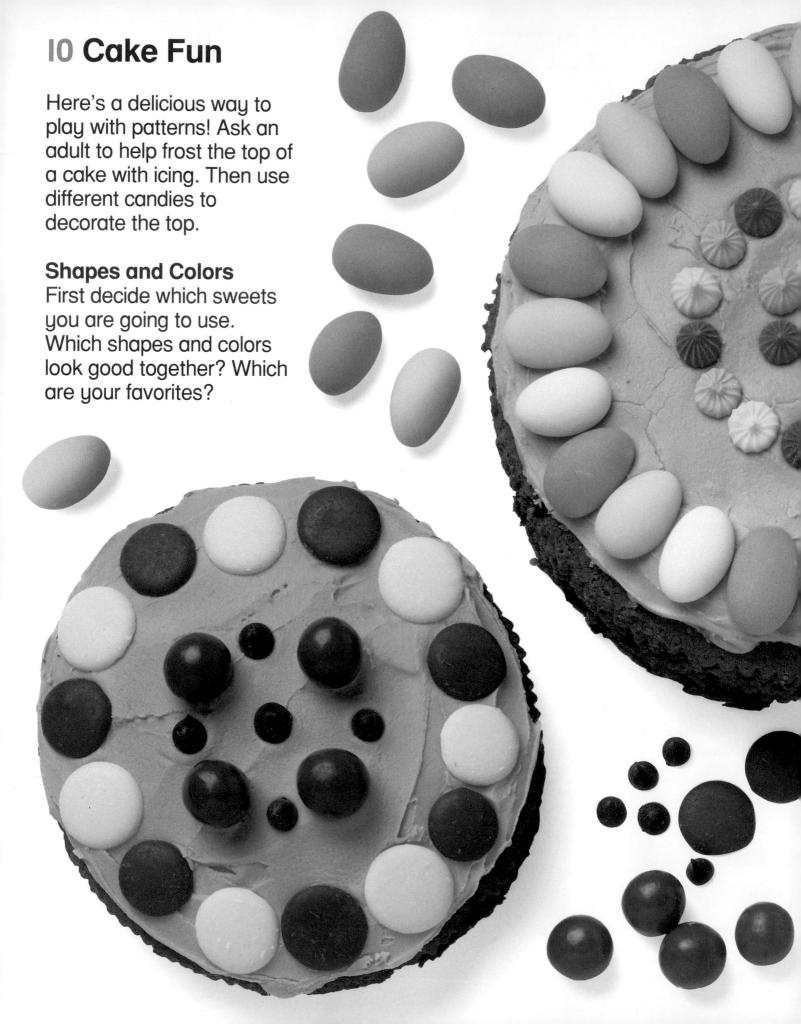

10 Cake Fun

Here's a delicious way to play with patterns! Ask an adult to help frost the top of a cake with icing. Then use different candies to decorate the top.

Shapes and Colors

First decide which sweets you are going to use. Which shapes and colors look good together? Which are your favorites?

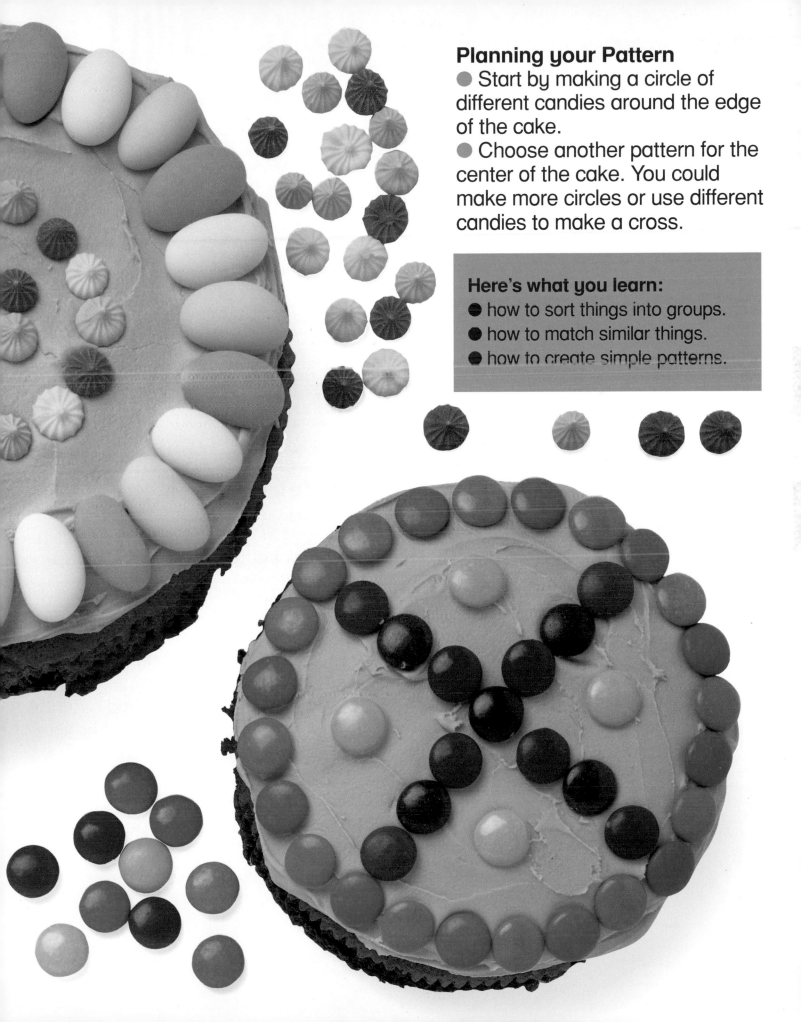

Planning your Pattern

● Start by making a circle of different candies around the edge of the cake.

● Choose another pattern for the center of the cake. You could make more circles or use different candies to make a cross.

Here's what you learn:
● how to sort things into groups.
● how to match similar things.
● how to create simple patterns.

12 Weaving

Some clothes are made from woven fabrics. The threads make a pattern. These fabrics are made on large machines called looms. You can do your own weaving at home using a cardboard loom and brightly colored yarn or strips of felt.

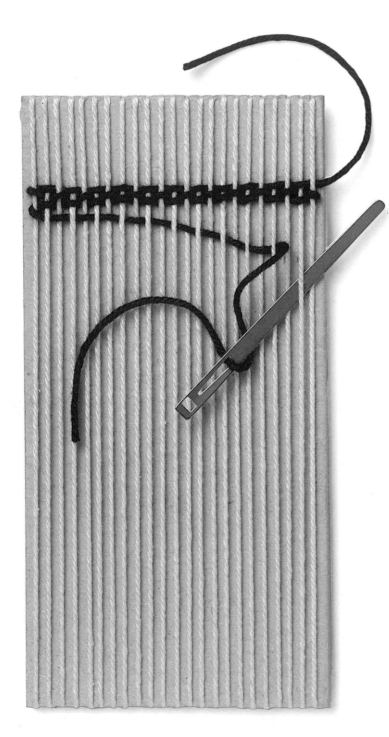

Loom
● Ask a grown-up to cut notches in two ends of a small piece of cardboard.
● Wind a piece of yarn around the cardboard. The notches will keep it in place. Tie the ends of the yarn at the back.

Over and Under
● Thread a large, blunt needle with a piece of yarn.
● Push the needle under and over the threads until you reach the other side.
● Weave back the other way, under the threads you went over before, and over those you went under.
● Cut the threads at the back to take your weaving off. Tie the ends together.

Here's what you learn:
● how to create simple patterns.
● how to use ideas about symmetry.

You can make a pattern with woven paper, too! First, find some fairly stiff colored paper.

Simple Pattern
● Fold a piece of paper in half.
● Make a row of cuts along the folded edge. Unfold the paper.
● Cut some strips of another color. Weave these strips through the slits as shown at right.

Diagonal Stripes
● Fold another piece of paper diagonally and make cuts. Weave strips as shown below or below right. Does the pattern look like the simple pattern?

Here's what you learn:
● how to create patterns and shapes.

Wavy Stripes
Cut wavy slits in a piece of paper.
You will also need some wavy strips
of another color. Weave the strips
as before.

Tartan Stripes
Cut two slits close together in a piece
of folded paper. Leave a gap, then cut
two more slits close together, and so
on. Weave thick and thin strips of
colored paper in and out of the slits.

Zig-zag Stripes
Ask an adult to make zig-zag slits in
the paper with a craft knife. Weave
straight strips through the slits.

16 Dot Patterns

Here's another way to make patterns with yarn.

Glue Patterns

● Make a pattern with a few dots of glue on a piece of poster board.
● Take a piece of yarn. Press one end into the glue.
● Guide the yarn around the glue pattern, pressing it down as you go.

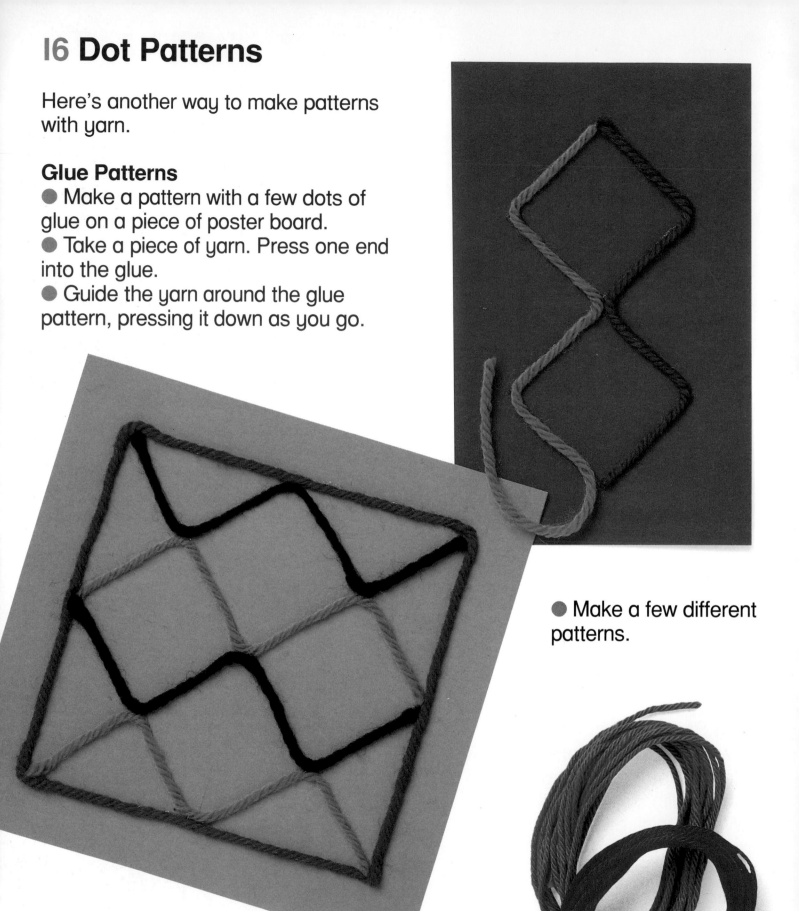

● Make a few different patterns.

Pin Patterns

● Arrange a pattern of pins on a piece of strong cardboard.

● Take a piece of yarn and tie it carefully to one of the outside pins.

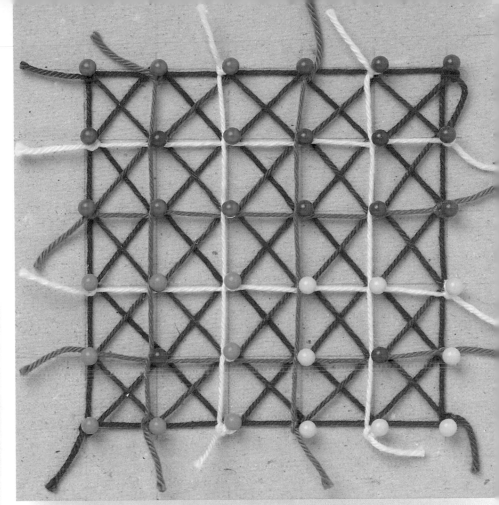

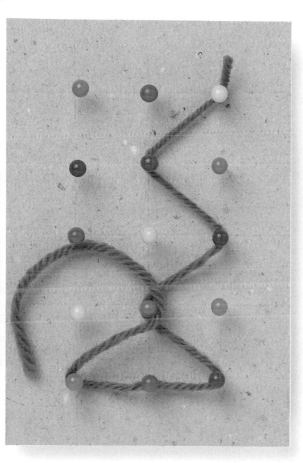

● Stretch the yarn around the pins, twisting it to keep it in place. When you reach the edge, tie the end of the yarn and cut off any extra.

Here's what you learn:
● how to create dot patterns.
● how to create shapes.

18 Cut Paper

You can make some amazing patterns by folding and cutting paper.

Fold and Cut

● Fold a piece of paper in half and then in half again. Cut a small piece out of one edge. Unfold the paper.

● To make a more complicated pattern, make several cuts along the edges before unfolding the paper.

Accordion Folds

● Cut a long strip of paper.
Fold it backward and forward,
so the strip is like an accordion.
● Ask an adult to help cut shapes out
of the folded paper.
● Unfold it to see your pattern.

Here's what you learn:
● how to create patterns.
● how to discover symmetrical patterns.

20 Tiles

Each of these tiles has a very simple design, but you can arrange them to make all kinds of patterns.

Designing the Tiles
● Cut out some squares of cardboard, all the same size.
● Choose a simple design and paint all the squares exactly the same.

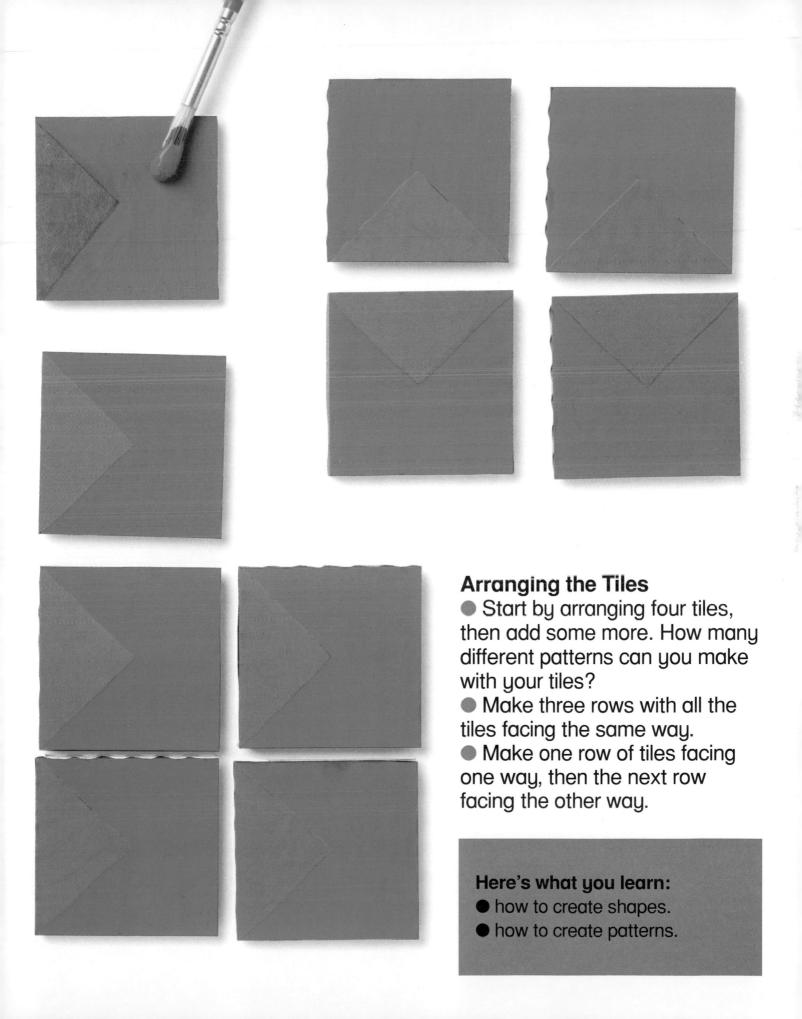

Arranging the Tiles

● Start by arranging four tiles, then add some more. How many different patterns can you make with your tiles?

● Make three rows with all the tiles facing the same way.

● Make one row of tiles facing one way, then the next row facing the other way.

Here's what you learn:
● how to create shapes.
● how to create patterns.

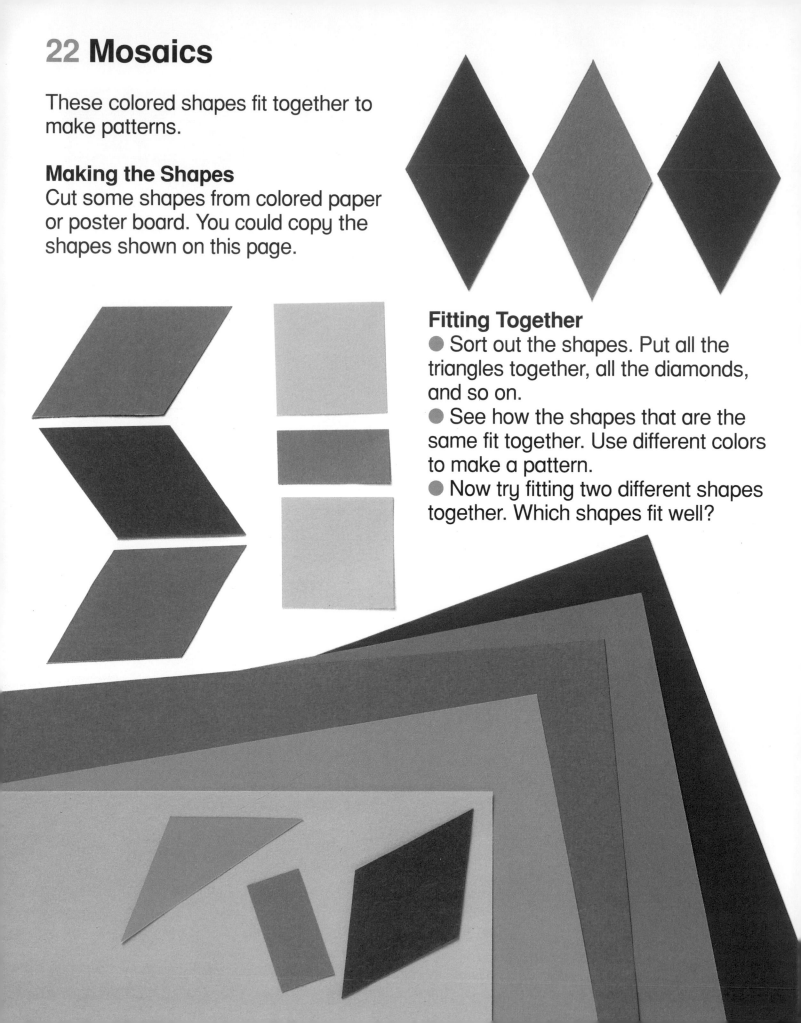

22 Mosaics

These colored shapes fit together to make patterns.

Making the Shapes
Cut some shapes from colored paper or poster board. You could copy the shapes shown on this page.

Fitting Together
● Sort out the shapes. Put all the triangles together, all the diamonds, and so on.
● See how the shapes that are the same fit together. Use different colors to make a pattern.
● Now try fitting two different shapes together. Which shapes fit well?

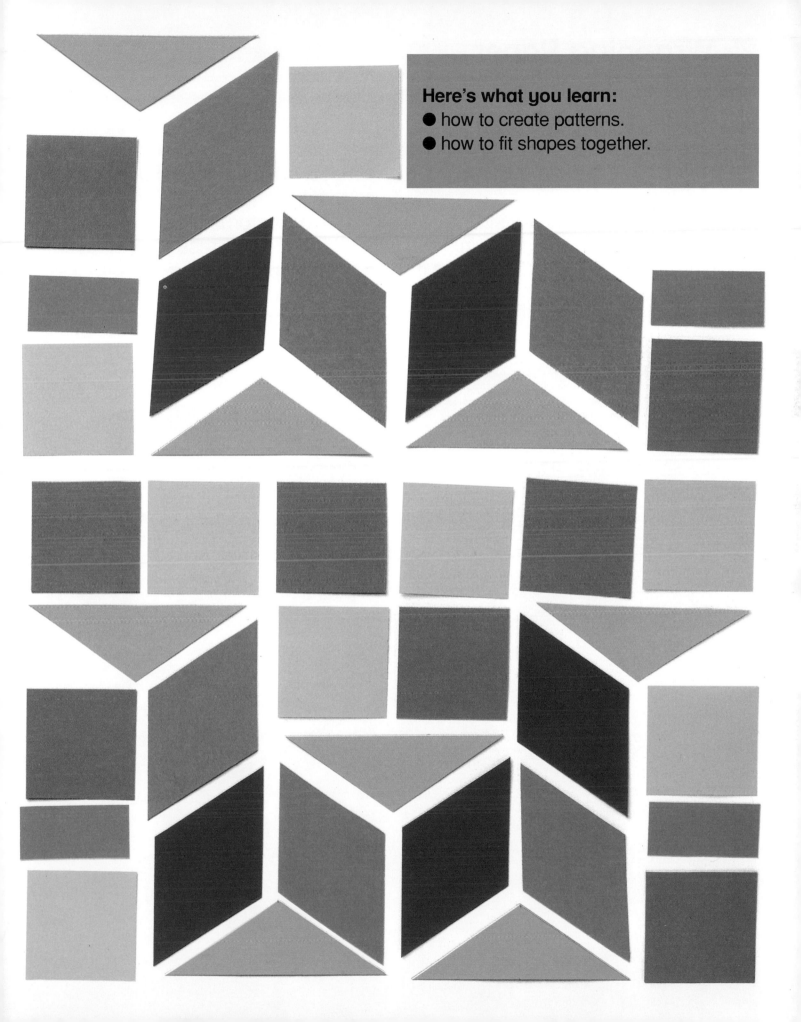

Here's what you learn:
● how to create patterns.
● how to fit shapes together.

24 Wrapping Paper

Make wrapping paper by decorating paper with a pattern. Pick a shape and repeat it lots of times.

String Blocks
● Glue a long piece of thick string onto a piece of cardboard and let it dry.
● Dip the string into paint and press it onto a sheet of paper.

Potato Prints
● Think of a simple shape and draw it on a piece of paper. Ask an adult to cut your shape from half of a potato so that the shape sticks up.
● Use a paintbrush to cover the shape with thick paint. Press the potato onto a sheet of paper. Then lift it off.
● Put some more paint on the potato. Print the shape lots of times.

Stencils

● Ask an adult to help cut a stencil out of poster board.

● Place the stencil on paper. Dab paint over the stencil. Remove it carefully and repeat.

Here's what you learn:
● how to create repeating patterns.
● how to invent new designs.

26 Mirror Prints

These amazing prints are reflections.

Fold and Paint
● Fold a piece of paper in half. Open it up and put a blob of paint on it.
● Fold the paper in half again and press down firmly. Then open it up.

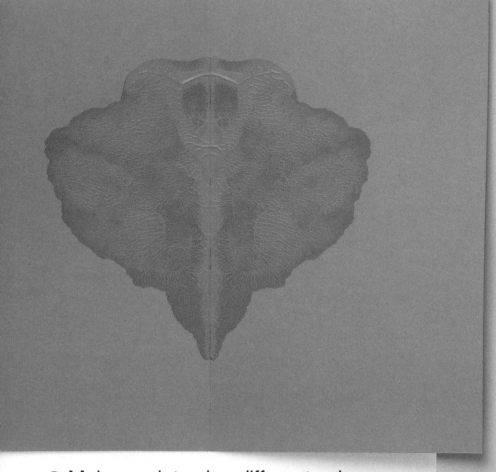

● Make a print using different colors. Allow one color to dry before adding the next.

Here's what you learn:
● how to use reflective symmetry.

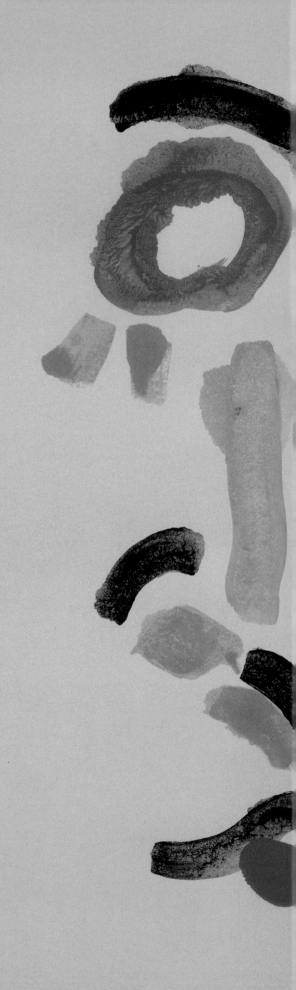

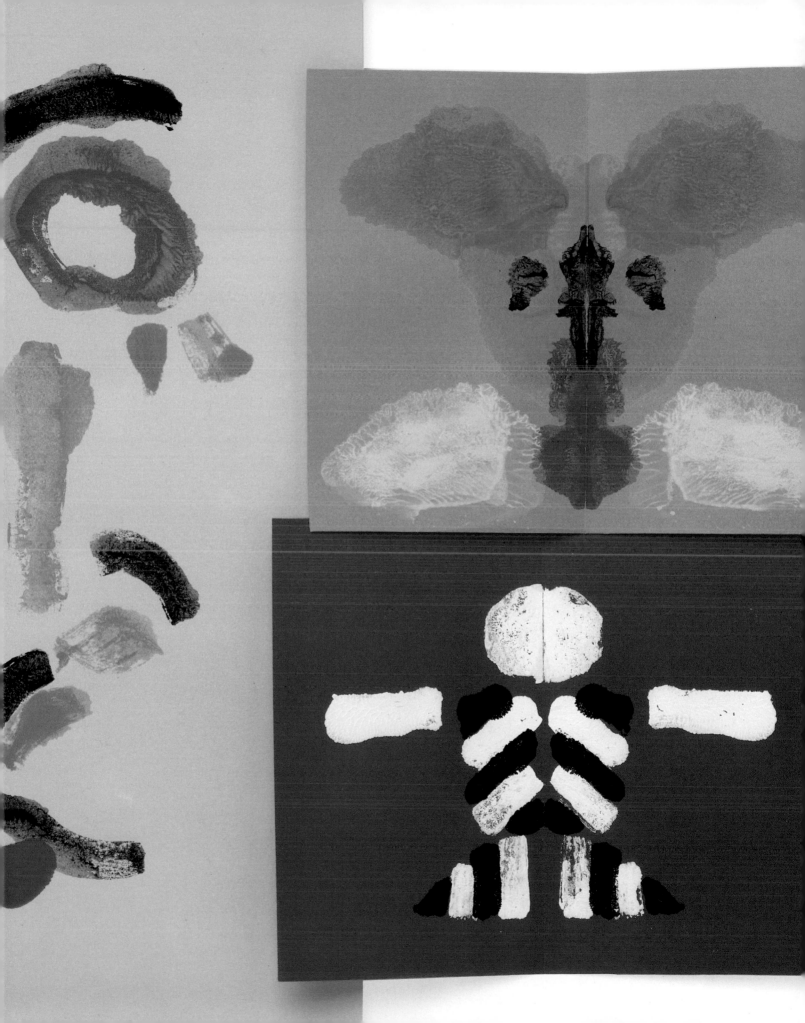

Think up some fun-looking patterns to use to decorate your T-shirts or socks. Then get permission to decorate some items. Use fabric paints, and make sure you read the instructions before you start.

Potato Patterns
● Ask an adult to cut a simple shape from half a potato.
● Cover the shape with paint, then press it onto your sock or T-shirt. Repeat to make a pattern.

Here's what you learn:
● how to create repeating patterns.
● how to use familiar shapes.
● how to invent new designs.

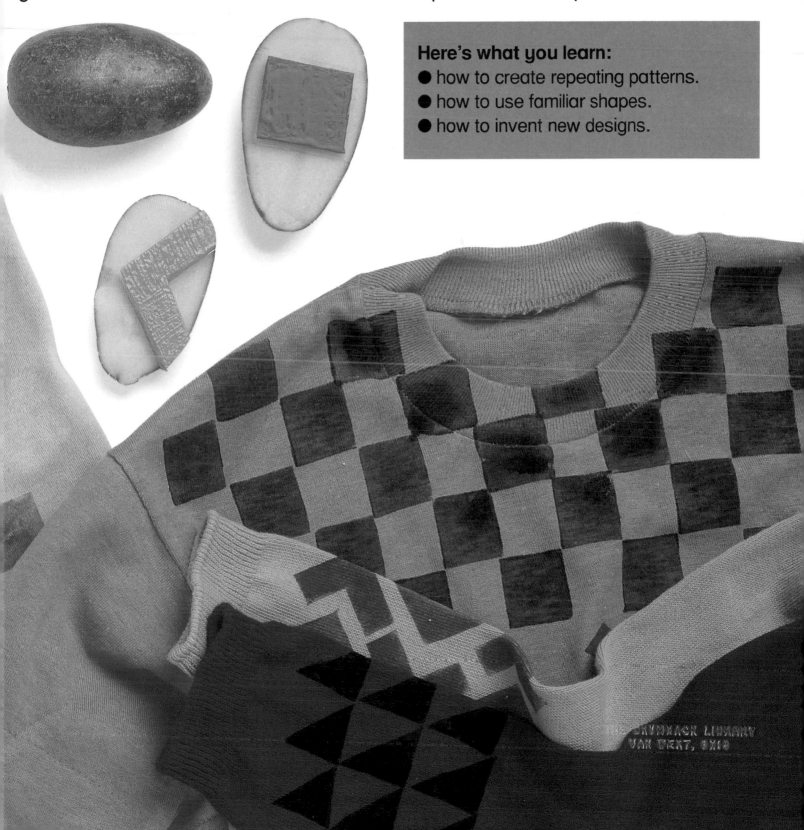

30 Stationery

Try using different patterns to decorate cards, writing paper, and envelopes. Use brightly colored paper and make sure you leave enough room to write! You could use some of the patterns you have found in this book or make up some new ones.

● Cut or tear shapes from colored paper. Glue them down in a pattern.
● Use plastic shapes to print.
● Cut out a cardboard stencil. Hold it down firmly and dab on paint with a sponge.

Here's what you learn:
● how to create patterns.
● how to create shapes.
● how to invent new designs.

Editor: Diane James
Photography: Toby
Text: Claire Watts

Published in the United States and Canada by
World Book, Inc.
525 W. Monroe Street
Chicago, IL
60661
in association with Two-Can Publishing Ltd.

**For information on other World Book products,
call 1-800-255-1750, x 2238,
or visit us at our Web site at http://www.worldbook.com**

Library of Congress Cataloging-in-Publication Data

Bulloch, Ivan.
    Patterns / Ivan Bulloch; consultants, Wendy and David Clemson.
        p.   cm. – (Action math)
    Originally published: New York: Thomson Learning, 1994.
    Includes index.
    Summary: Teaches the skills of pattern recognition, sorting,
matching, and pattern creation by means of various handicrafts.
    ISBN 0-7166-4902-0 (hardcover)—ISBN 0-7166-4903-9 (softcover)
    1. Geometry–Juvenile literature.   2. Handicraft–Juvenile
literature.   [1. Geometry. 2. Handicraft.]   I. Clemson, Wendy.
II. Clemson, David.   III. Title.   IV. Series: Bulloch, Ivan.   Action
math.
QA445.5.B85   1997
516'.15–dc21   96-49559

Printed in Hong Kong

2 3 4 5 6 7 8 9 10 01 00 99 98 97

Skills Index

Consultants
Wendy and David Clemson are
experienced teachers and
researchers. They have written
many successful books on
mathematics.